Rodrigo Alavarse

Radiomodifying evaluation of folic acid in Danio rerio embryos

Rodrigo Alavarse

Radiomodifying evaluation of folic acid in Danio rerio embryos

Exposed to gamma radiation

ScienciaScripts

Imprint
Any brand names and product names mentioned in this book are subject to trademark, brand or patent protection and are trademarks or registered trademarks of their respective holders. The use of brand names, product names, common names, trade names, product descriptions etc. even without a particular marking in this work is in no way to be construed to mean that such names may be regarded as unrestricted in respect of trademark and brand protection legislation and could thus be used by anyone.

Cover image: www.ingimage.com

This book is a translation from the original published under ISBN 978-3-639-68778-1.

Publisher:
Sciencia Scripts
is a trademark of
Dodo Books Indian Ocean Ltd. and OmniScriptum S.R.L publishing group

120 High Road, East Finchley, London, N2 9ED, United Kingdom
Str. Armeneasca 28/1, office 1, Chisinau MD-2012, Republic of Moldova, Europe
Managing Directors: Ieva Konstantinova, Victoria Ursu
info@omniscriptum.com

Printed at: see last page
ISBN: 978-620-8-58285-2

I dedicate this work to my mother, Aparecida Delfonso Alavarse,

thank you for everything.

ACKNOWLEDGEMENTS

First of all, I thank the Eternal, YHWH, the Word who became flesh and dwelt among us, who allowed everything to happen.

I would like to thank Dr José Roberto Rogero and Sizue Ota Rogero for their welcome, trust and opportunity.

Special thanks to Dr Daniel Perez Vieira, for his help and commitment in the CEBIO laboratory, and Murilo Álison Vigilato Rodrigues, CEQMA laboratory, for sharing his knowledge.

Thanks to Aurea Silveira Cruz from the Adolfo Lutz Institute for her contribution to the research.

I would especially like to thank the CEQMA staff and students, Susanna Busch, Silvia Rejane, Eliane de Santana, Clóvis Souza Santos, Esther Fracaro, Patrícia S. da Silva and Diego Mendonça, who helped directly and indirectly.

I would like to thank Elizabeth Sebastiana Ribeiro Somessari from CETER for her excellent work, help and understanding.

I would like to thank Dr Raúl Bonne Hernández, UNIFESP, for his help during difficult times.

I would like to thank Dr Martha Simões Ribeiro for her tips, Carolina dos Santos Moreno and Juliana de Souza Azevedo for their participation in the examining board.

I would like to thank IPEN for the opportunity and the infrastructure provided for the research and CAPES for the financial support.

"Blessed are you who can show that co-operation is better than fighting or competing. That way, you'll discover who you really are and where you belong. that they occupy in the family of God."

Matthew 5:9

SUMMARY

ALAVARSE, R. D. ***Radiomodifying evaluation of folic acid in Danio rerio embryos exposed to gamma radiation****. 2024. 46 p. Dissertation (Master's Degree in Nuclear Technology), Nuclear and Energy Research Institute, IPEN-CNEN/SP, São Paulo.*

The study investigated the potential of folic acid (FA) as a radiomodifying agent in zebrafish embryos exposed to ionising radiation (IR). FA, a B-complex micronutrient, acts as a coenzyme in amino acid metabolism and in DNA and RNA synthesis. Using cell viability assays, the median inhibitory concentration (IC50) of FA was determined in the NCTC-L929 cell line (mouse connective tissue) (2.366 mg/mL). Tests were then carried out to determine the median lethal dose (LD50) of IR in embryos irradiated at 4, 6 and 24 hours post-fertilisation (hpf) (LD50, 4 hpf - 96h: 9.77 Gy and 24 hpf - 144 h: 12.85 Gy). Subsequently, median lethal concentration (CL50) tests were carried out on zebrafish embryos exposed for 120 hours (CL50: 0.679 mg/mL). The results suggest that FA has radioprotective potential, standing out as a possible therapeutic strategy, providing important contributions for future research. Organisms irradiated after incubation with HF showed significant survival, with 96 hpf increasing by 12%, 17% for 120 hpf and 22% at 144 hpf.

Keywords: *Folic acid, Zebrafish, Radiomodifiers, Gamma radiation*

SUMMARY

1 INTRODUCTION

The estimate for new cases of cancer, according to data from the National Cancer Institute (INCA) for the three-year period 2023-2025, is 704,000 new cases, excluding non-melanoma skin cancer. Figure 1 shows the 10 types of cancer with the highest incidence for both sexes. Consequently, the search for radiotherapy treatments is increasing and, in specific cases, the use of radiomodifying agents is necessary (CHAL et al., 2020).

Figure 1: Estimate for the 10 most incident cancer cases for both sexes

Localização Primária	Casos	%	Homens	Mulheres	Localização Primária	Casos	%
Próstata	71.730	30,0%			Mama feminina	73.610	30,1%
Cólon e reto	21.970	9,2%			Cólon e reto	23.660	9,7%
Traqueia, brônquio e pulmão	18.020	7,5%			Colo do útero	17.010	7,0%
Estômago	13.340	5,6%			Traqueia, brônquio e pulmão	14.540	6,0%
Cavidade oral	10.900	4,6%			Glândula tireoide	14.160	5,8%
Esôfago	8.200	3,4%			Estômago	8.140	3,3%
Bexiga	7.870	3,3%			Corpo do útero	7.840	3,2%
Laringe	6.570	2,7%			Ovário	7.310	3,0%
Linfoma não Hodgkin	6.420	2,7%			Pâncreas	5.690	2,3%
Fígado	6.390	2,7%			Linfoma não Hodgkin	5.620	2,3%

Source: Inca, 2023.

Ionising radiation (IR) is widely used in industry, health and research. In healthcare, the use of IR can range from simple diagnostics to complex radiotherapy treatments. Radiotherapy, for example, uses IR beams to destroy neoplastic cells, generally X-rays, gamma rays and electron beams are the most commonly used types (BONTRAGER et al., 2010).

The effects of IR can vary depending on the type of molecule affected (water, protein, DNA, RNA) and the damage can be caused by direct ionisation of macromolecules or by reactive oxygen species resulting from water ionisation by-products (NOUAILHETAS et al., 2005). These reactions can promote cell death, genetic mutations or be repaired by the irradiated organism (BUSHONG et al., 2010).Folic acid (FA) is a B-complex micronutrient, vitamin B9, which acts as a coenzyme in the metabolisation of amino acids, the formation of purines and pyrimidines for the synthesis of DNA and RNA, essential for cell division. O balanced intake of PA is essential for homeostasis, as it plays an important role in nucleic acid synthesis and regeneration in methylation reactions (DE MARQUIA et al., 2015).Insufficient intake of PA or its poor absorption by the body allows metabolic abnormalities such as hyperhomocysteine and DNA hypomethylation to appear, consequently leading to chronic diseases and dysfunction in foetal development (BAILEY and GREGORY III, 1999).PA supplementation in cases of deficiency is generally administered to patients with megaloblastic anaemia resulting from cancer treatments (VANNUCCHI and MONTEIRO, 2010). The Danio rerio or zebrafish is a freshwater fish used as an animal model in research related to human pathologies, ecotoxicity, medications and other chemical substances. Its genome contains characteristics similar to those of humans, including DNA repair (SCHNEIDER et al., 2009).

Embryonic development takes place rapidly and important organs can be observed within 48 hours post-fertilisation (hpf), making genetic manipulation simple. Research into the viability and development of zebrafish embryos exposed to gamma radiation has shown that they are susceptible to its effects, and it is possible to observe changes in cellular organisation and morphology, enabling comparison with clinical therapies (DAMASCENO, 2019).

The effects of IR on the radiomodifying action of FA are not clearly known, so further research is extremely necessary to provide results on its action in exposed organisms.

2 OBJECTIVES

The aim of this research is to evaluate the radiomodifying action of folic acid on Danio rerio embryos.

Specific objectives:

- Carry the folic acid cell viability test to obtain the 50% inhibitory concentration (IC50);
- Obtain 50% lethal dose (LD50) values for gamma radiation in Danio rerio embryos;
- Obtain 50% lethal concentration (LC50) values of folic acid in embryos of Danio rerio;
- Investigating the radiomodifying action of folic acid on Danio rerio embryos exposed to gamma radiation on the basis of previous tests.

3 REVIEW BIBLIOGRAPHY

3.1 History of folic acid

Between the 1920s and 1930s, Lucy Wills described a new haematopoietic element in yeast and liver that could cure a type of tropical anaemia in India, called "vitamin M". By the early 1940s, it was common knowledge that eating liver could cure pernicious anaemia, and the first isolation of the substance in a study evaluating growth factors place. Stokstad and Manning discovered growth factors in chicken meat, which they called "Factor U", while Hogan and Parrot identified anti-anaemic factors in chicken liver extract, which they called "Vitamin BC". In mid-1945, Angier and colleagues described the synthesis of pteroylmonoglutamic acid (VANNUCCHI and MONTEIRO, 2010). Wills and his colleagues later discovered other factors present in dark green leaves that stimulated the growth of Streptococcus lactis, which was named folic acid. The term folic acid, folium (leaf) from Latin, is due to its presence in vegetable leaves. Spies reported that FA cured megaloblastic anaemia during pregnancy (VANNUCCHI and MONTEIRO, 2010). As early as 1962, diagnostic parameters for PA deficiency were established and criteria for need in adults were established, which serve as the basis for current supplementation recommendations. Still in the first half of the 20th century, with the identification of PA synthesis, nutritional treatments were prescribed for anaemias, and later on there was an emphasis on the between absorption pathways and deficiency with congenital cardiovascular pathologies and cancer. Since 1990, research has shown that high levels of homocysteine are associated with neural tube defects during pregnancy. In order to reduce anomalies in the foetus, the American

Public Health Service advises a daily intake of PA in women of reproductive age (VANNUCCHI and MONTEIRO, 2010). In 2002, in Brazil, Anvisa published Public Consultation No. 51, which became Collegiate Board Resolution No. 344, requiring fortification of wheat and maize flours and their by-products with FA, where for every 100 g of a product it must contain a minimum of 0.15 mg of FA. The prevention of neural tube pathologies in the foetus, myelomeningocele and anaemia were the main reasons for this requirement in the country. In 2017, a new resolution was published in order to avoid compounds with low bioavailability (BRASIL, 2002; 2017).

3.1.1 Chemical Structure of Folic Acid

PA is derived from exogenous sources and is a compound with the molecular formula C19H19N7O6. The most stable form is pteroylglutamic acid, which is not metabolically active. The molecular formula includes any member of the pteroylglutamate family, has a conjunction pteroic acid and at least one L-glutamic acid residue (Figure 2). It is generally found in the form of tetrahydrofolate, the active form of the vitamin, which is a monoglutamate, resulting from the reduction of pteroylglutamic acid, starting from substitutions in positions 5 and 10 of the pteridine nucleus, metabolised by the liver (VANNUCCHI and MONTEIRO, 2010).

Figure 2: Chemical structure folic acid

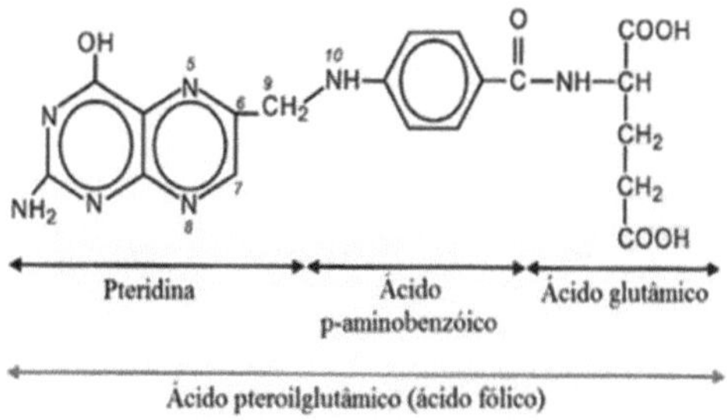

Source: Vannucchi, 2010.

3.1.2 Sources of folic acid

The foods with the highest PA content are yeast, dark green leafy vegetables, liver and other offal, peanuts, eggs, whole grains and other enriched cereals (VANNUCCHI and MONTEIRO, 2010).

3.1.3 Danio rerio

Danio rerio (zebrafish), commonly known as paulistinha, is a teleost fish belonging to the cyprinid class, native to South Asia, whose adults grow to between 3 and 5 cm, with an average weight of 1.5g. Its diet is omnivorous, consuming a range of benthic and planktonic crustaceans. It is very resistant to temperature variations (18°C to 30°C), with an ideal of 24°C to 28°C and tolerates a pH range of 6.0 to 8.0, the ideal being 6.4 to 7.4 (SANTOS, 2012). Sexual dimorphism can be affected by environmental factors, although this separation can be made in the laboratory by observing certain characteristics. Male fish have straighter bodies, shorter lengths and more intense colours than females (Figure

3). (SANTOS, 2012; DAMMSKI, 2011).

Figure 3: Zebrafish couple, the male (left) has a more rectilinear body compared to the female (right).

Source: Zebrafish, Manual de criação em biotério.

Because it is a small fish, it can be farmed in large quantities in a small space, with low maintenance costs. They reach reproductive age between 3-6 months, the females spawn daily and in suitable water conditions they produce larger numbers of embryos per spawning, on average 500 per day (ZON, 1999; DAMMSKI, 2011). Embryogenesis occurs very quickly and as a result, important organs such as the brain, eyes, heart, kidneys and gastrointestinal tract can be seen within 48 hours of fertilisation (hpf) (GEIGER et al., 2006). Figure 4 shows the main stages of zebrafish development.

Figure 4: Zebrafish development stages: (0 hpf) is the moment of spawning, passing through the stages of cleavage (4 hpf), gastrulation and epiboly (6-8 hpf), hatching (3-4 days), until reaching the adult stage in 90 days.

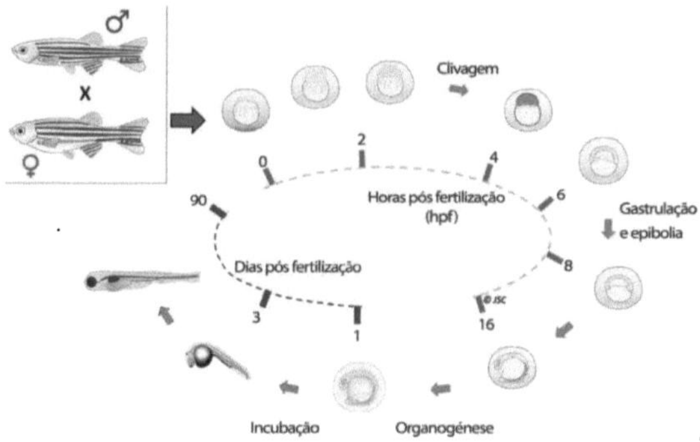

Source: Casa das Ciências.

In toxicological assessments, there is identification of lethality factors or toxicity indicators and their mechanisms that facilitate understanding of the dose-response relationship and toxicodynamics of the substance tested. There are many data morphology and physiological information at all stages development (HILL et al., 2005). Use at the embryonic stage allows the organisms to be manipulated in cell culture microplates, allowing for several replicates at once and a greater variety of toxicity results. The organisms survive for several days by obtaining nutrients from the yolk sac, so there is no need to feed them. They also have transparency, which allows the organism's development to be observed under a microscope (Figure 5). (HILL et al., 2005; GEIGER et al., 2006).

Figure 5: Zebrafish embryo 24 hpf. transparency allows all development to be observed.

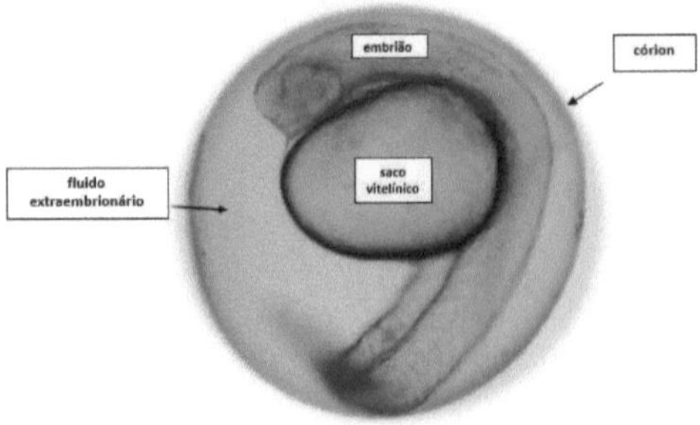

Source: adapted from Diário jove.

3.1.4 Radiation ionising

IR can be propagated in the form of waves or particles and has sufficient energy to remove electrons from atoms or molecules in the medium through which it propagates. IR emission occurs through the natural or artificial decay of unstable nuclei, or through the excitation of atoms and their nuclei in nuclear reactors, X-ray tubes, cyclotrons and other equipment (UNITED NATIONS, 2000). Ionising sources usually have their greatest use and technological development in medicine for the diagnosis and treatment of cancer. Many of these use electron beams, X-rays and gamma rays (BUSHONG et al., 2010). Despite technological advances in radiotherapy procedures, the surrounding tissues in the treatment target area end up suffering adverse reactions, which compromise patients' quality of life and the treatment itself, which in some cases must be interrupted (GREENBERGER, 2009).

3.1.5 Radiomodifiers

The use of chemical substances that promote protection against the harmful effects of IR was further developed after the Second World War. The effect of the amino acid cysteine on rats exposed to radiation was a pioneering study. These substances are extremely important and can be applied beyond clinical practice to workers and even emergency plans (DAMASCENO, 2019).These substances are known as radiomodifiers and can act as a radioprotector with antioxidant properties, helping the body to combat harmful effects and repair macromolecules by capturing free radicals before they damage the cell. Or they can act as radiosensitisers, potentiating the harmful effects of IR (SANTOS, 2013). The literature shows that vitamins A, E and C, flavonoids and phenolic compounds have antioxidant action. Various types of plant extract have been tested as radioprotectants, demonstrating a positive effect against IR damage, but there is still a need to elucidate the mechanisms of action of these compounds (DAMASCENO, 2019).

4 MATERIALS AND METHODS

4.1 Substance and organism test

O AF, with a molar mass of 441.4 g/mol, was purchased from Sigma-Aldrich in the form of a dark yellow powder with 99% purity. The adult fish used for reproduction were provided by Companhia Ambiental do Estado de São Paulo (CETESB) and reared in the ecotoxicology laboratory of the Chemistry and Environment Centre (CEQMA) of the Nuclear and Energy Research Institute (IPEN). The use of animals for this research was approved and registered by the Ethics Committee on the Use of Animals n.249/19. The aquariums were maintained using reconstituted water, pH, conductivity and temperature parameters following ABNT - 2007 standards, which establish methods for testing fish, including short-term chronic toxicity. Specific guidelines for zebrafish established by OECD 236 - Fish Embryo Toxicity (FET), which is used to assess the toxicity of chemical substances in fish embryos in order to determine environmental impacts, identify and prevent ecological risk and comply with regulations, were applied to carry out the tests on embryos. For spawning, the male and female organisms remained in separate aquariums for a period of 7 to 14 days. The day before, they were placed in specific breeding tanks (Figure 6) and kept out of the light. In the morning the lights are switched on, simulating daylight, mating then takes place and the eggs are laid at the bottom of the breeding aquarium (ABNT, 2007).

Figure 6: Aquariums used for breeding zebrafish.

Source: the author.

Figure 7 shows the scheme for selecting the embryos and placing them in 24-well plates for acute toxicity tests.

Figure 7: Embryo selection scheme.

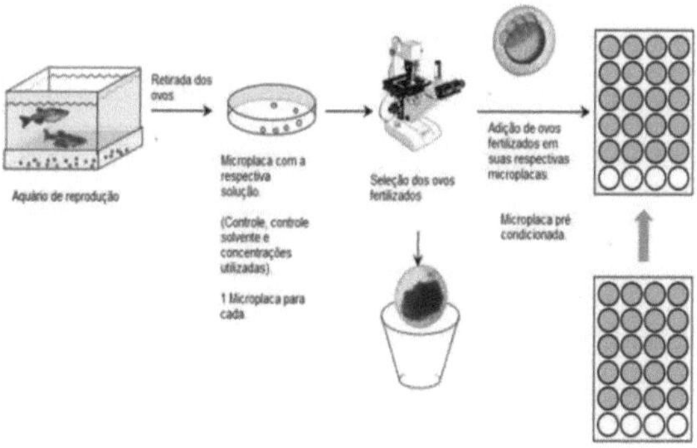

Source: adapted from OECD 236

4.2 Minimum concentration required to inhibit 50% of cells (IC_{50})

The tests were carried out on NCTC-L929 mouse connective tissue cells adhered in 96-well plates supplied by the Cell Culture Centre of the Adolfo Lutz Institute, SP, Brazil. Cell viability was checked using supravital staining of a tetrazolium compound (MTS). The MTS dye is reduced in the same way as MTT, to formazan compound in the presence of viable cells. The formazan produced can be measured in a spectrophotometer at an absorbance of 490 nm (FRESGAHNEY, 2015). In this study, four types of solvent were tested with FA: dimethylsulphoxide (DMSO); sodium bicarbonate (NaHCO3 0.1M); sodium hydroxide (NaOH 1M) and phosphate-buffered saline (PBS). The microplates with the cell culture were filled with the respective initial solutions of AF 4 mg/mL, sequentially with six more (1; 0.5; 0.25; 0.125; 0.0625; 0.03125 mg/mL), using minimum essential medium (MEM) for dilution.The microplates were incubated at 37°C in a humid atmosphere with 5% CO2 for 24 hours. The wells were then washed and the MTS dye was added and left to incubate for a further three hours before going into the reader. From these tests, the solution that showed the best solubility for the FA was chosen, taking into account the dilution time, the lower cytotoxicity of the vehicle and the greater amount of diluted FA. Cell viability was read using a Thermo Multiskan EX spectrophotometer (Figure 8) with a wavelength of 490 nm. The results obtained were collated and evaluated using the Prism GraphPad version 10.0.0 statistical programme, using the "Dose-response-inhibition curves" equation.

Figure 8: Thermo Multiskan EX microplate reader

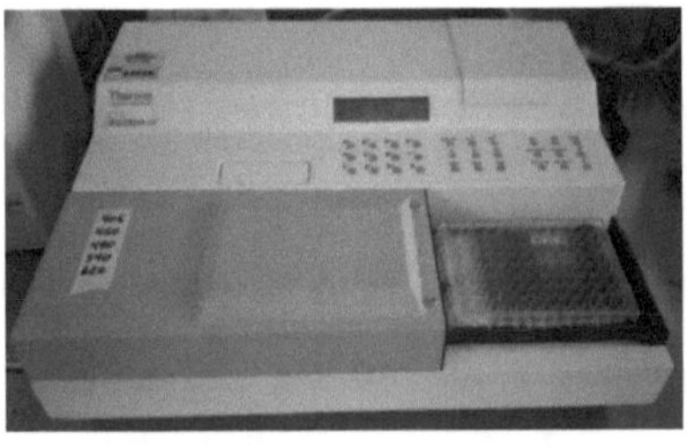

Source: The author.

4.3 50% lethal dose of gamma radiation on zebrafish embryos (LD_{50})

The tests to obtain the LD50 of gamma radiation in zebrafish embryos had validity criteria based on OECD protocol no. 236, where only batches with a fertilisation rate of ≥70%, temperature 26±1°C, control survival rate ≥ 90%, hatching rate 80% were used (OECD, 236).

In five 24-well microplates, only the embryonic medium (ultrapure water and salts that enable the embryo to live) was used to fill in where the embryos would be placed. The plate controls for irradiated embryos were made separately, with embryos from the same batch, with only 1 embryo per well. The embryos were irradiated at three different times of development, 4, 6 and 24 hpf. (SANTIAGO, 2021)

For each test, the embryos were irradiated at five different doses, 5, 10, 15, 20 and 25 Gy, for 4 and 6 hpf. Embryos at 24 hpf were irradiated with higher doses of 10, 20, 30, 40 and 50 Gy, as previous studies have shown that at this stage of development the organism is more resistant (DAMASCENO, 2019). The organisms were irradiated in the Gammacell cobalt-60 source (Figure 9) with a dose rate of 392.23 Gy/h^{-1}, located at IPEN's Radiation Technology Centre (CETER).

The tests lasted a total of 144 hours. The organisms were observed daily using an optical microscope (Olympus CK40). LD50 values were obtained at 96, 120 and 144 hpf. Embryo mortality and anomalies were counted and the data added to the ORIGIN 8.0 statistical programme, using a dose-response function to obtain the LD50.

Figure 9: Gammacell 220, with ^{60}Co source.

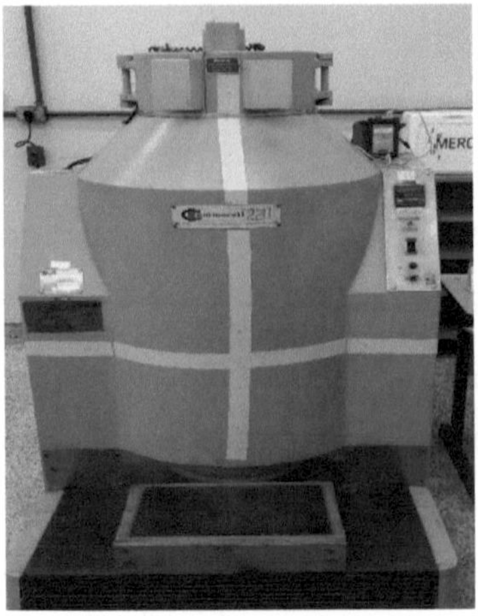

Source: The author.

4.4 Acute toxicity of folic acid in Danio rerio embryos (CL_{50})

Based on the IC50 obtained in the cytotoxicity test, the PA concentration was chosen for the 50% lethal concentration test in zebrafish embryos (CL50). The entire test was based on OECD protocol no. 236. All the requirements for the validity of the test are in accordance with the standards described for obtaining the LD50. The diluted PA molecule is highly hydrophilic and uses a genetically distinct and functionally diverse transport system to enter cells. For this experiment, 1 g of FA was diluted in 9 mL of PBS + 1 mL of NaoH and then embryonic medium was added, forming 1 litre, pH=7.4 (mother solution) (BELLAVINHA, 2012). The yolk sac of the embryos is highly permeable, so the PA molecules will be absorbed for metabolisation. Figure 10 shows embryos at different stages of development (DE ALMEIDA et al., 2022).

Figure 10: Zebrafish embryos at different stages of development: A= 2 hpf, B= 11.5 hpf, C= 24 hpf.

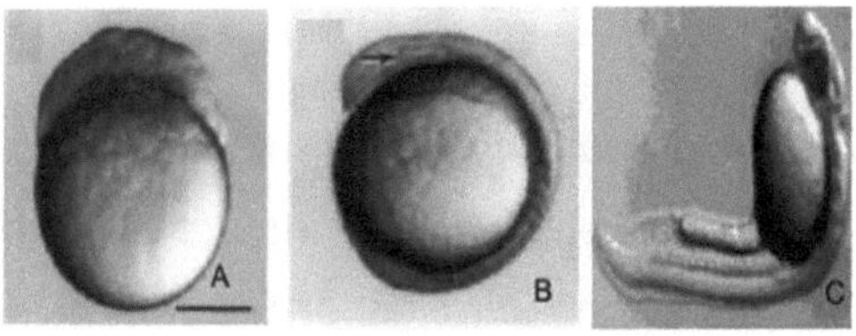

Source: zfin.org

Table 1 shows the time of embryo development, the sequential dilutions for each plate and the amount of PA (mg/mL) dissolved in the embryo medium, carried out in triplicate.

Table 1: Embryos selected at 2 hpf, dilution sequence (%) and amount of AF (mg/mL), 100% corresponds to the mother solution.

Hours post fertilisation (hpf)	Dilutions (%)	Folic acid (mg/mL)
2	100	1
	50	0,5
	25	0,25
	12,5	0,125
	6,25	0,0625
	3,125	0,03125

Source: The author.

The embryos remained in the HF solutions throughout the test period and were observed every 24 hours, for a total duration of 144 hours. CL50 values were at 96, 120 and 144 hpf. Mortality and changes in development were noted and counted in these processes. The data was added to the ORIGIN 8.0 statistical programme, using the dose-response function to obtain the CL50.

4.5 Evaluation of the radiomodifying effect of folic acid on embryos from zebrafish

To determine the radiomodifying effect of HF, viable embryos were selected immediately after spawning and incubated in HF solution in petri dishes for 3 hours at a concentration of 0.5 mg/mL (Figure 11). This solution was prepared in the same way as the previous test to determine the CL50, only the percentage of PA was modified.

The embryos were then separated into irradiated groups with and without the presence of PA, as well as non-irradiated groups with and without the presence of PA. Plate controls were duly established to check the quality of the batch of embryos on other plates, since it is not feasible to protect the plate control during the irradiation process.

Three different batches were used for the three tests carried out, each containing 20 embryos, one per microplate well. After incubation, the organisms were washed twice in embryonic medium to remove excess PA and transferred to 24-well plates containing only embryonic medium, then irradiated with 4 hpf and kept in an incubator. throughout the test period. The organisms were observed at 96, 120 and 144 hpf, and the data was collected and analysed in Origin 8 using the dose-response function. A statistical analysis was carried out to check for significant differences in the survival of organisms irradiated with and without the presence of PA.

Figure 11: Zebrafish embryos 3 hpf incubated with AF.

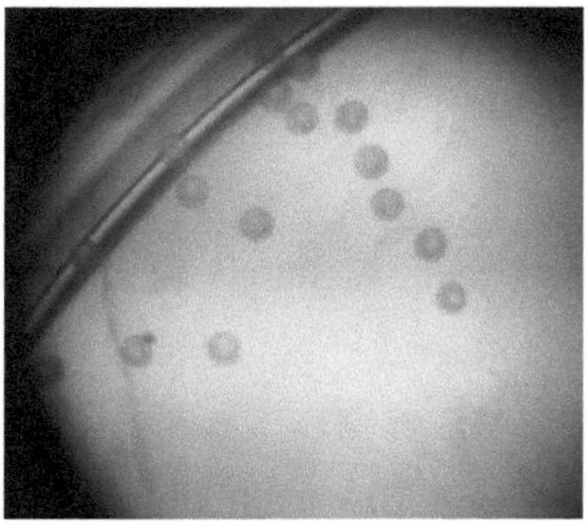

Source: The author.

5 RESULTS AND DISCUSSION

5.1 Minimum concentration required to inhibit 50% of cells (IC_{50})

Folic acid is often used in cell culture to evaluate its effects on cell proliferation, differentiation and metabolism. However, specific studies combining pure FA with MTS or MTT assays in cell culture, similar to this study, are not very common.The IC50 obtained in the AF+DMSO test was 0.523 mg/mL. The cell viability curves are shown in Figure 12. DMSO presented limitations for the tests, because when the proportion of 4 mL and 4 mg of AF was increased, the solution became pasty, making it difficult to handle. Increasing the amount of DMSO also had a cumulative effect on PA toxicity. Because of these difficulties, it was discarded for use in tests with zebrafish embryos.

Figure 12: Viability of NCTC-L929 cells in AF+DMSO solution (IC50= 0.523 mg/mL, DMSO concentration 7.04%).

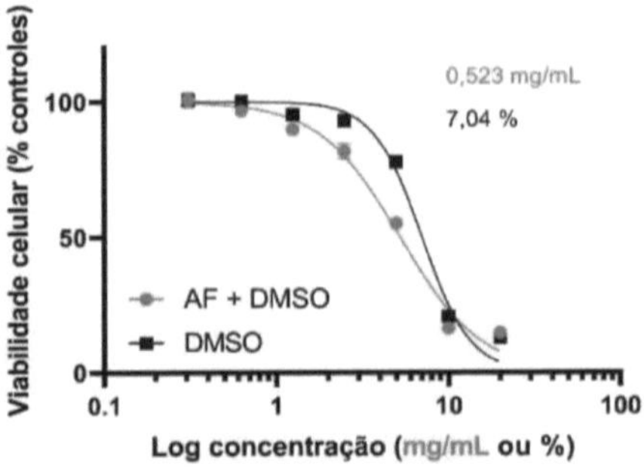

Source: the author.

The cell viability of AF+NaOH showed an IC50 value of= 0.668 mg/mL, corresponding to 3.69 % of the vehicle (Figure 13). For AF+NaHCO3, the

IC50 value was 2.471 mg/mL, corresponding to a low toxicity of the vehicle> 100% (Figure 14).

Figure 13: Viability of NCTC-L929 cells in AF+NaOH solution (IC50= 0.668 mg/mL, NaOH concentration 3.687%).

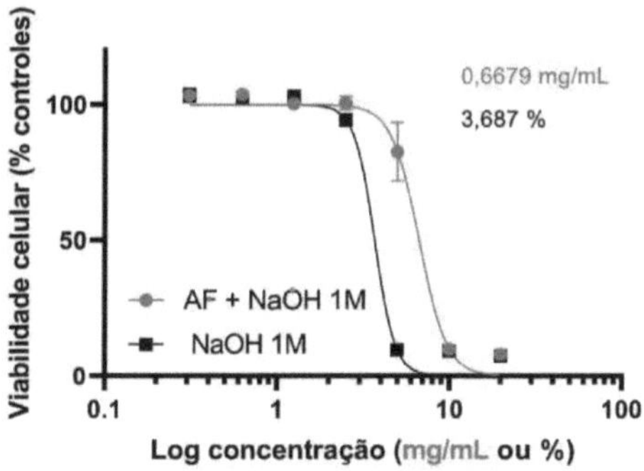

Source: the author.

Figure 14: Viability of NCTC-L929 cells in AF+NaHCO3 solution (IC50= 2.471 mg/mL, NaHCO3 concentration >100 %).

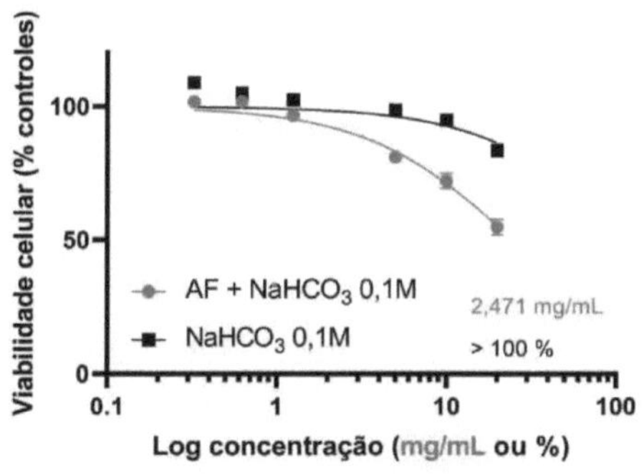

Source: the author.

The cell viability of AF+PBS indicated an IC50 value of 2.366 mg/mL, which corresponds to 49.18% of the vehicle (Figure 15).

Figure 15: Viability of NCTC-L929 cells in AF+PBS solution (IC50= 2.366 mg/mL, DMSO concentration 49.18%).

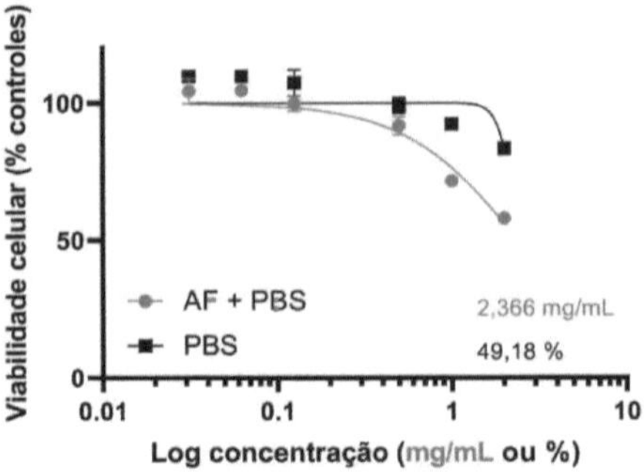

Source: the author

The IC50 values of the AF+NaHCO3 and AF+PBS solutions showed similar results, although NaHCO3 alone showed lower toxicity, PBS was better suited to mixing with embryonic medium for the zebrafish test, the pH correction factor (7.4) was decisive, Santos (2024), who evaluated the radiomodifying effect of folic acid in vitro, also chose PBS as the solvent for AF in his study.

5.2 50% lethal dose of gamma radiation on zebrafish embryos (LD_{50})

The acute toxicity tests to obtain the LD50 in organisms irradiated with 4 hpf are shown in Table 2. Table 3 shows the LD50 values in organisms irradiated with 6 hpf.

Table 2: DL50 values obtained for organisms irradiated with 4 hpf

Horas pós Fertilização (hpf)	Tempo após irradiação (h)	DL_{50} (Gy)
4	96	9,77±0,48
	120	8,71±1,28
	144	6,75±0,15

Source: the author.

Table 3: DL50 values obtained for organisms irradiated with 6 hpf

Horas pós Fertilização (hpf)	Tempo após irradiação (h)	DL_{50} (Gy)
6	96	14,68±0,56
	120	12,31±1,78
	144	11,84±0,25

Source: the author.

Table 4 shows the LD50 values for organisms exposed for 24 hpf, which could not be calculated for the 96 hpf period due to the low lethality.

Table 4: DL50 values obtained for organisms irradiated 24 hpf

Horas pós Fertilização (hpf)	Tempo após irradiação (h)	DL_{50} (Gy)
24	96	-
	120	16,17±1,09
	144	12,85±1,16

Source: the author.

In this study, a significant difference was observed in the LD50 values in embryos of organisms irradiated at 4 and 6 hpf. Organisms irradiated with 4 hpf at early stages of development showed greater sensitivity to the effects of gamma radiation. At doses above 10 Gy, greater mortality was observed in organisms with 4 hpf compared to 6 hpf. This pattern was more evident with doses above 15 Gy, indicating greater resistance in organisms irradiated with 6 hpf. Organisms irradiated at 24 hpf reached LD50 values after 120 hpf, especially with doses above 15 Gy, which suggests that the stage embryo development should be taken into account when observing the effects of ionising radiation. These results are in line with data in the literature. Kumar et al. (2017), observed an increase in the mortality rate in embryos exposed to gamma radiation as the dose increased. Damasceno (2018), in his study with zebrafish embryos using the same radiation doses as this study, used a specific compound to remove the chorion from the embryos and concluded that there was no difference in the lethality of organisms exposed with and without the chorion. The results obtained in this study are sufficient to determine the radiation doses to be used in the next stages, where the embryos will be irradiated in the presence and absence of PA.

5.3 Acute toxicity of folic acid in Danio rerio embryos (CL_{50})

Determining the CL50 for organisms at 96 hpf was not feasible due to the low lethality rate observed. , at 120 hpf the CL50 was 0.679 mg/mL (Figure 16), while at 144 hpf, the CL50 obtained was 0.544 mg/mL (Figure 17).

Figure 16: Lethality of zebrafish embryos immersed in PA solution (CL50= 0.679 mg/mL,

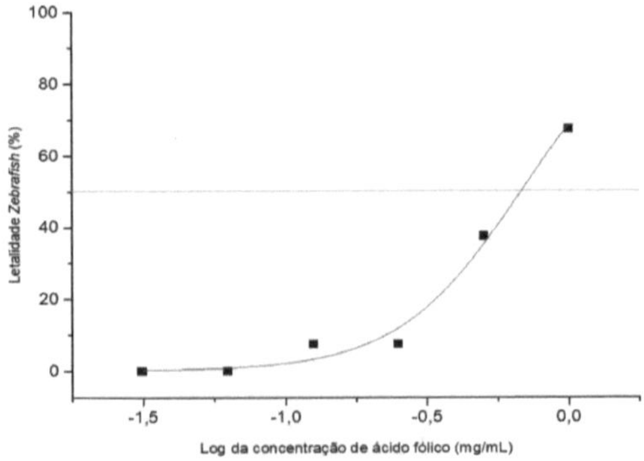

Source: the author.

Figure 17: Lethality of zebrafish embryos immersed in PA solution (CL50= 0.544 mg/mL

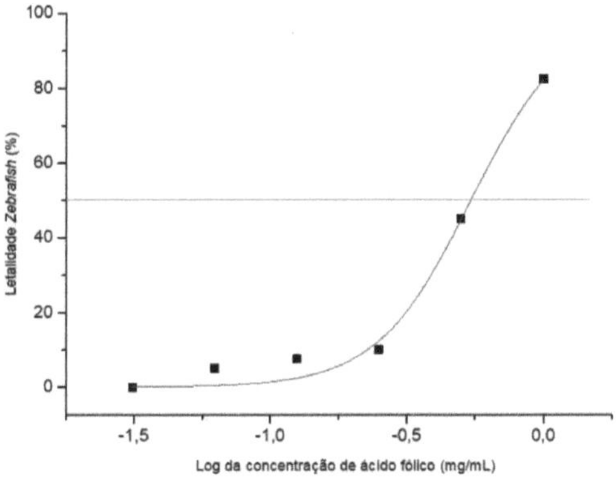

Source: the author.

The solutions with the highest concentrations of PA showed a positive correlation with lethality at 120 and 144 hpf, although determining the CL50 value was not feasible for organisms at 96 hpf, as there was a reduction in the hatching rate and the occurrence of oedema in the pericardium of the embryos at this stage.Nascimento (2022), in his study on nutrition in zebrafish breeding, emphasises the importance of folic acid as one of the main vitamins for development, although there are still no recommended values for its supplementation. It was observed that organisms exposed to concentrations of less than 50% in this study did not show significant damage compared to the control group. MA et al. (2012) evaluated the preventive effect of PA on zebrafish embryos to prevent sodium selenite toxicity, observing an improvement in embryo survival and a reduction in cases of pericardial oedema.

5.4 Evaluation of the radiomodifying effect of folic acid on zebrafish embryos

The groups exposed to a radiation dose of 10 Gy showed a lower survival rate compared to the groups irradiated with the same dose and incubated with HF. Figure 18 illustrates a significant difference at 96 hpf (12 per cent). The survival rate decreases over time, as seen in Figure 19, where greater mortality is evident in the irradiated organisms after 120 hpf (17%), compared to those only irradiated. This observation becomes more evident in Figure 20, 144 hpf (22%).

Figure 18: Zebrafish survival 96 hpf, organisms irradiated with 10 Gy compared to organisms irradiated with 10 Gy + AF.

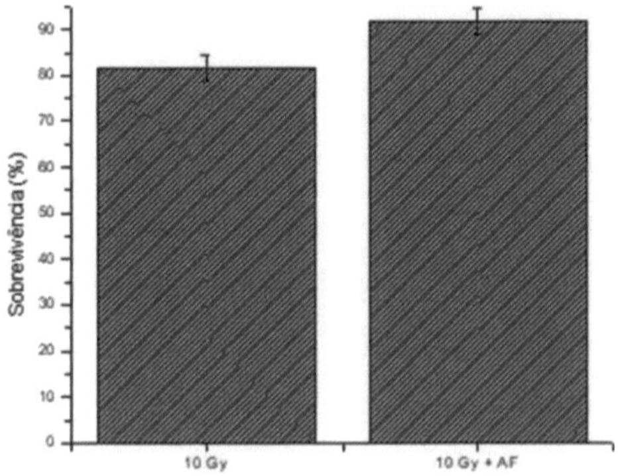

Source: the author.

Figure 19: Zebrafish survival 120 hpf, organisms irradiated with 10 Gy compared to organisms irradiated with 10 Gy + AF.

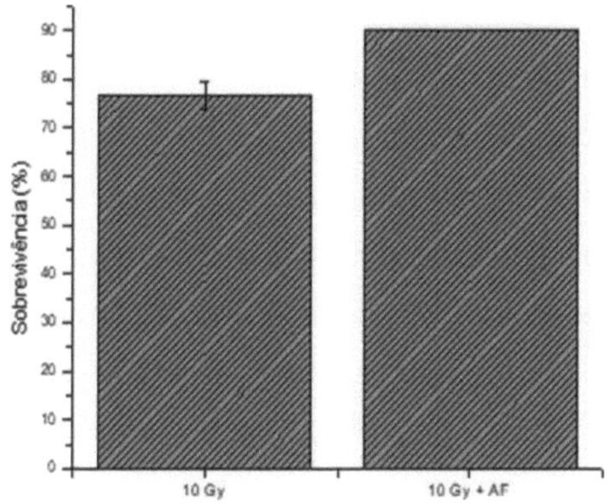

Source: the author.

Figure 20: Zebrafish survival 144 hpf, organisms irradiated with 10 Gy compared to organisms irradiated with 10 Gy + AF.

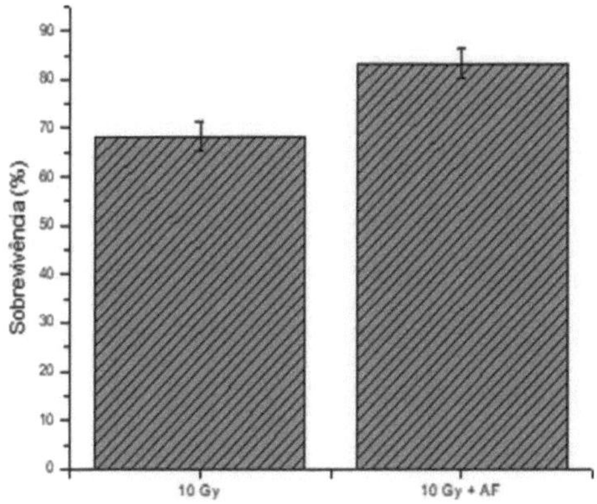

Source: the author.

Padula and Ponzinibbio et al. (2016) investigated the effects of PA supplementation on protection against DNA damage and chromosomal instability induced by low-dose IR, using Chinese hamster ovary (CHO) cells as an in vitro model. The results revealed that PA supplementation reduced apoptosis, DNA damage and chromosomal instability caused by low-dose IR. PA supplementation may play a protective role against the adverse effects of radiation at low levels. This highlights the importance of PA status when investigating the effects of radiation and emphasises the need for further research to better understand the effects of PA supplementation in different radiation exposures. The study by Hussein et al. (2018), investigated the potential of magnetic nanoparticles loaded with FA as an approach to mitigate the adverse effects of gamma radiation in rats. The results showed that the combination of these nanoparticles with FA can help protect against DNA damage, reduce IR-induced triglyceride and cholesterol levels, and improve the antioxidant system, preserving tissue integrity, especially in the liver. These findings suggest that this intervention may be promising in the prevention of harmful effects, opening doors to possible applications in radioprotection and tissue preservation.Craenen et al. (2021) studied the effects of prenatal exposure to X-radiation on rat foetuses and the preventive potential of PA supplementation. The results revealed that PA was able to partially prevent severe defects, such as anophthalmia, exencephaly and agnathy, induced by IR. In addition, PA showed benefits in reducing prenatal mortality, foetal weight and axial skeletal defects. However, some defects, such as iris anomalies, open eye and gastroschisis, did not respond PA supplementation.

The study also examined the effects of irradiation on the morphology and function of the adult nervous system, showing a decrease in brain volume and visual acuity, as well as alterations in olfactory performance, which were partially alleviated. The mechanisms underlying the radioprotection of FA include its antioxidant role, DNA repair, DNA methylation and transcriptional regulation. It also highlights the potential of FA as a strategy to protect foetuses against the harmful effects of IR, pointing to the need for further research to evaluate its efficacy in humans and determine the optimal dose of supplementation.Radiation exposure can cause adverse effects in developing zebrafish embryos, such as the formation of scoliosis, a curvature or twist in the . This is due to damage to the DNA developing cells, which can lead to cell death or mutations that affect the proper formation of tissues. At critical stages of development, radiation interferes with normal processes such as cell differentiation and migration, which are essential for the correct formation of the spine. In addition, radiation induces oxidative stress in cells, resulting in further damage to DNA and developing tissues. This can deregulate the expression of genes involved in spinal formation and musculoskeletal development, resulting in structural defects (MACEDA and VISCOR, 2015).Figure 21 shows the relationship between the observation of spinal deviation in embryos irradiated with 10 Gy with and without the presence of HF. Despite observing a lower incidence in organisms that were incubated with HF, the result was not significantly significant in of radioprotection, for a better analysis of the results of these deformities observed would be A focused study using a greater number of specific tests and analyses is needed. The incidences of deformities presented in this study complement the radiomodified assessment of PA and open up possibilities for future research.

Figure 21: Incidence of spinal deviation in embryos irradiated with 10 Gy with and without the presence of AF.

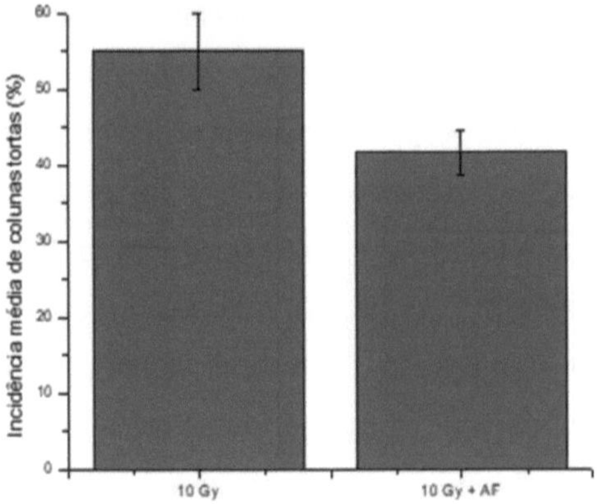

Source: the author.

Figure 22 shows embryos in a normal state and the column deviation.

Figure 22: Column deviation in irradiated zebrafish larvae, 96 hpf. a- Normal column and b- column deviation.

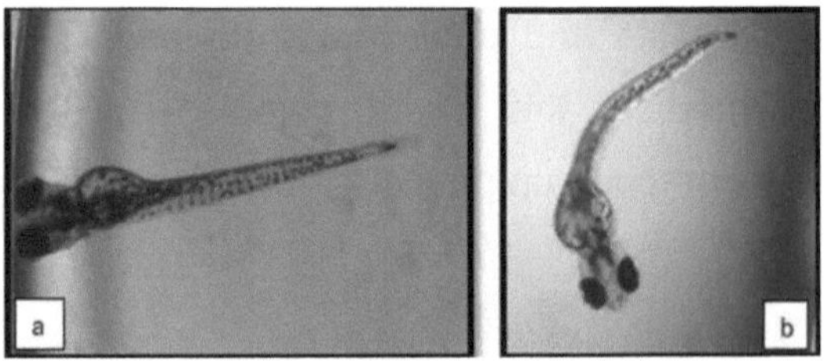

Source: the author.

Al-enezi et al. (2014), observed that exposure to gamma radiation negatively affected embryonic development and embryo growth. Malformations and growth delays in embryos exposed to gamma radiation, indicating that radiation may have adverse effects on the early development of these fish. Organisms exposed to high doses of IR generally show a delay in hatching or do not hatch at all. When comparing the dose effect in organisms irradiated with 10 Gy with and without incubation with HF, no significant differences in hatching were observed, as shown in figure 23.

Figure 23: Incidence of zebrafish hatching irradiated with 10 Gy with and without incubation with AF.

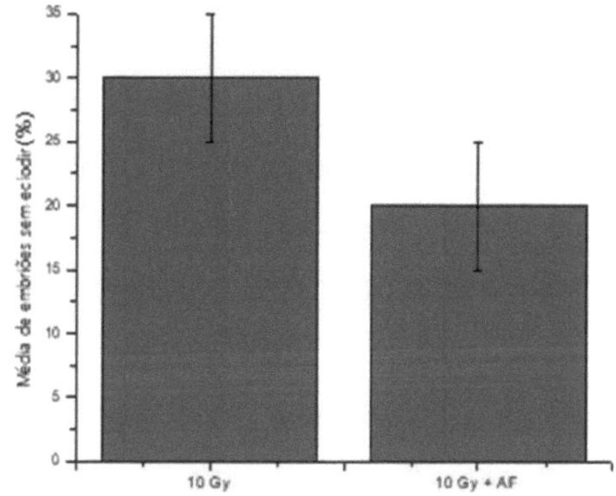

Source: the author.

Figure 24 shows embryos at different hpf that did not hatch.

Figure 24: Embryos that failed to hatch, on the left an embryo irradiated with a dose of 10 Gy after 96 hpf, on the right an embryo irradiated with a dose of 10 Gy after 120 hpf and previously incubated with AF.

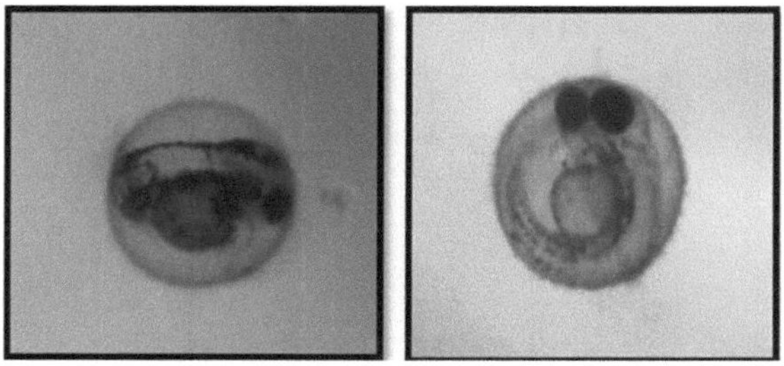

Source: the author.

Studies using zebrafish embryos subjected to irradiation have been conducted to elucidate the effects of radiation exposure on brain morphology and development. These investigations aim to assess the impacts of different doses of radiation on brain organogenesis and determine whether such effects can result in neuropathological conditions such as microcephaly (KAMSTRA, 2018).Spinal deviation and failure to hatch were the main effects found in the development of the organisms in this study. Compared to the effects of skull deformity and the action of FA, there were also no significant differences, as shown in figure 25. Figure 26 shows zebrafish larvae belonging to the control group and another with cranial reduction, suggestive of microcephaly.

Figure 25: Incidence of skull deformity in zebrafish irradiated with 10 Gy with and without incubation with HF.

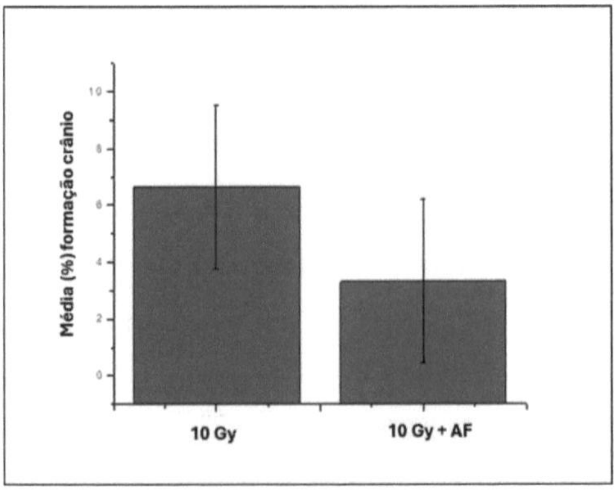

Source: the author.

Figure 26: Zebrafish larvae at 96 hpf, on the right belongs to the control group showing normal development characteristics and on the left another subjected to incubation with PA and exposure to radiation.

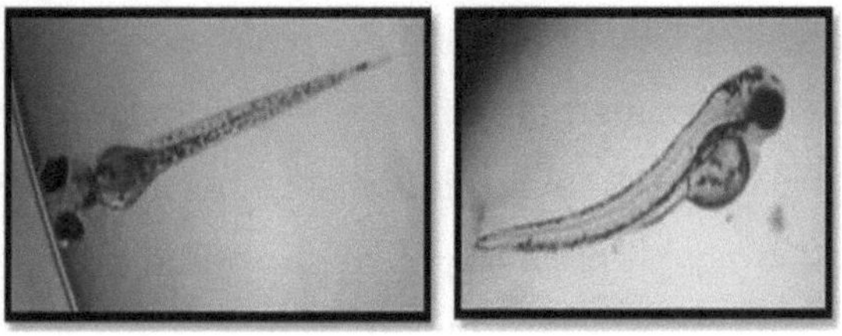

Source: the author.

Precioza and Mita (2020) observed in their study that exposure to X-radiation resulted in abnormalities in cardiac development, such as pericardial oedema. The occurrence of oedema in the pericardium also showed no significant difference when comparing organisms irradiated with and without incubation with HF, as shown in Figure 27.

Figure 27: Incidence of pericardial oedema in zebrafish irradiated with 10 Gy with and without incubation with HF.

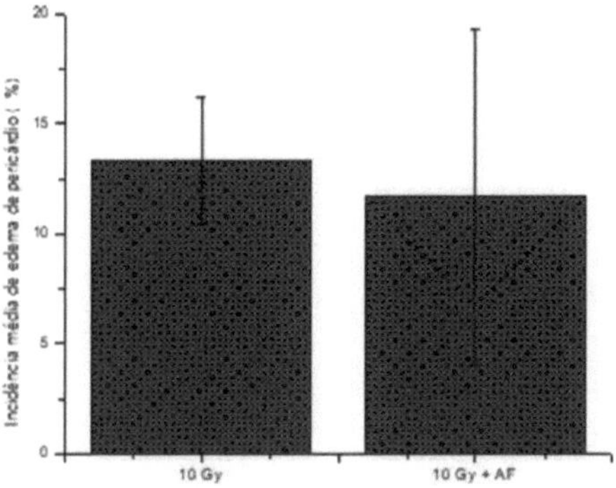

Source: the author.

6 CONCLUSIONS

The IC50 value of AF in PBS was determined to be 2.366 mg/mL, maintaining a stable pH of 7.4 in the embryonic medium.The LD50 values in embryos irradiated at 4, 6 and 24 hpf varied with development time.The CL50 values of PA in zebrafish embryos at 120 hpf= 0.679 mg/mL and 0.544 mg/mL at 144 hpf. PA showed a radioprotective effect in irradiated organisms after incubation, with greater survival, at 96 hpf it increased by 12%, 17% for 120 hpf and 22 per cent at 144 hpf.

7 REFERENCES BIBLIOGRAPHIC

AL-ENEZI, E.; AL-BADER, D.; AL-DHFYAN, A. et al. The effect of gamma radiation on early development and growth of zebrafish (Danio rerio). **International Journal of Radiation Biology**, v. 90, n. 2, p. 112-116, 2014.

BRAZILIAN ASSOCIATION OF TECHNICAL STANDARDS (ABNT). **Brazilian standard regulation no. 15.499: aquatic ecotoxicology - short-term chronic toxicity test method with fish**. 2007. Available at: https://ojs.brazilianjournals.com.br/ojs/index.php/BRJD/article/download/4611/427 6/12496. Accessed on: 10 January 2024.

BAILEY, Lynn B.; GREGORY III, Jesse F. Folate metabolism and requirements.

The Journal of Nutrition, v. 129, n. 4, p. 779-782, 1999. BELLAVINHA, Karime Rezende. **Evaluation of folic acid as a potential candidate for biosafety**. 2012.

BONTRAGER, Kenneth L.; LAMPIGNANO, John P. **Treatise on radiographic positioning and associated anatomy**. Rio de Janeiro: Elsevier Brasil, 2010. 784 p.

BRAZIL. National Health Surveillance Agency (ANVISA). Collegiate Board Resolution - RDC no. 344, of 13 November 2002. Provides for the control of medicines subject to special control and makes other provisions. **Diário Oficial da União**, Brasília, DF, 14 Nov. 2002. Section 1, p. 76-78. Available at: ANVISA. Accessed on: 11/09/2024.

BRAZIL. National Health Surveillance Agency (ANVISA). Collegiate Board Resolution - RDC no. 166, of 23 July 2017. Provides for the enrichment of foods with vitamins and minerals. **Federal Official Gazette**, Brasília, DF, 24 July 2017. Section 1, p. 38-47. Available at: ANVISA. Accessed on: [11/09/2024].

BUSHONG, Stewart. **Radiological science for technologists: physics, biology and protection**. Rio de Janeiro: Elsevier, 2010. 1015 p.

CHAL, Juliana Lopes et al. Ionizing radiation radiomodifying vegetables. **Acta Scientiae et Technicae**,v.8,n.1,p.1-33,2020.Available at: https://www.researchgate.net/publication/343578815_Vegetais_Radiomodificador es_de_Radiacoes_Ionizantes. Accessed on: 22 August 2023.

CRAENEN, Kai et al. **Folic acid fortification prevents morphological and behavioural consequences of X-ray exposure during neurulation. Frontiers in Behavioural Neuroscience**, v. 14, p. 609660, 2021.

DAMASCENO, K. **Radiomodifying evaluation of resveratrol Danio rerio embryos irradiated with gamma radiation.** 2019. Dissertation (Master's Degree in Nuclear Technology) - Institute for Nuclear and Energy Research, IPEN.

CNEN/SP, São Paulo. Available at: www.teses.usp.br. Accessed on: 25 May 2019.

DAMMSKI, Ana Paula; MÜLLER, Bruno Roberto; GAYA, Carolina. **Animal Husbandry Manual**. 2011.

DE ALMEIDA, Thais Victoria Pires et al**. Toxicity of sediments contaminated by mining tailings in the Doce River on zebrafish embryos (Danio rerio)**. 2022.

DE MARQUIA, Paola Agatina et al. Main factors of low adherence to folic acid use. **Journal of Health Sciences**, v. 16, n. 2, 2015.

FRESHNEY, R. Ian. Animal cell culture: manual of basic technique and specialised applications. **John Wiley & Sons**, 2015.

GEIGER, G.A. et al. **Zebrafish as a "biosensor"? Effects of ionising radiation and amifostine on embryonic viability and development.**

Cancer Research, v. 66, n. 16, p. 8172-8181, 2006.
GREENBERGER, J.S. Radioprotection. In Vivo, v. 23, n. 2, p. 323-336, 2009. HILL, A.J. et al. Zebrafish as a model vertebrate for investigating chemical toxicity. **Toxicological Sciences**, v. 86, n. 1, p. 6-19, 2005.
HUSSEIN, Samy Ali et al. Biochemical role of folic acid loaded magnetic nanoparticles versus gamma irradiation in rats. Benha Veterinary **Medical Journal**,v. 34, n. 3, p. 14-25, 2018.
KAMSTRA, Jorke H. et al. Ionising radiation induces transgenerational DNA methylation effects in zebrafish. **Scientific Reports**, v. 8, n. 1, p. 15373, 2018.
KUMAR, M.K.P. et al. Effects of gamma radiation on the early developmental stages of zebrafish (Danio rerio). **Ecotoxicology and Environmental Safety**, v. 142, p. 95-101, 2017.
MA, Yan et al. Embryonic developmental toxicity of selenite in zebrafish (Danio rerio) and prevention with folic acid. **Food and Chemical Toxicology**, v. 50, n. 8,p. 2854-2863, 2012.
MACEDA-VEIGA, A.; GREEN, A. J.; DE SOSTOA, A.; VISCOR, G. Embryonic exposure to ionising radiation causes developmental arrest and dysregulation of early neurogenesis in the zebrafish (Danio rerio). **PLOS ONE**, v. 10, n. 12, e0143596, 2015.
NASCIMENTO, Carolina Bulhões do et al. **Important aspects zebrafish (Danio rerio) nutrition in vivariums**. 2022.
NOUAILHETAS, Yannick et al. Ionising radiation and life. Rio de Janeiro: **CNEN**, 2005. 22 p.
OECD. **Guideline for the testing of chemicals. Fish embryo acute toxicity (FET) test**, n. 236, July 2013.

PADULA, Gisel; PONZINIBBIO, María Virginia; SEOANE, Analia Isabel. **Possible radioprotective effect of folic acid supplementation on genomic instability induced by low-dose ionising radiation in vitro**.

2016.

PREZIOSA, G.; RIZZO, F.; CARMOSINO, M.; GIORDANO, F.; MITA, D. G. Effects of X-ray radiation on the early development of zebrafish (Danio rerio) embryos.
PLOS ONE, v. 15, n. 7, e0235774, 2020.

SANTIAGO, Magda Regina. **Analysing the effects of pesticides based on imidacloprid and abamectin on zebrafish embryos (Danio rerio)**. 2021. PhD Thesis - University of São Paulo.

SANTOS, C. S. **Study of the radiomodifying effect of folic acid in vitro**. 2024. Dissertation (Master's Degree in Nuclear Technology) - Nuclear and Energy Research Institute, IPEN-CENEN SP, São Paulo.

SANTOS, D.C. **Endocrine disruption in zebrafish (Danio rerio) embryos, analyses of gene expression and embryonic development**. 2012. 132 f. Dissertation (Master's Degree in Biotechnology) - University of Trás-os-Montes e Alto Douro, Portugal.

SANTOS, Felipe Tiago José dos. **Evaluation of the radiomodifying effect of lichenic substances**. 2013. Dissertation (Masters Degree) - Federal University of Pernambuco, Postgraduate Programme in Energy and Nuclear Technologies, Recife.

UNITED NATIONS. Scientific Committee on the Effects of Atomic Radiation. Sources and effects of ionising radiation: sources**. United Nations Publications**, 2000.

VANNUCHI, Helio; MONTEIRO, Thaís Helena. **Folic acid. Fully recognised functions of nutrients**. Translation. São Paulo.

MIX
Papier aus verantwortungsvollen Quellen
Paper from responsible sources
FSC® C105338

Printed by Books on Demand GmbH, Norderstedt / Germany